Questioning

the

Star of Life

Acknowledgements

This book is dedicated to everyone who has interest in space and stars, or likes to question the existence of other lifeforms. Thank you for reading.

Sources: -YOUTUBE Channel
        SpaceRip. Are there other earths: The odds of life around nearby stars documentary-
        -Space.com
    How many stars are in the Milky Way? -How many stars are in the universe?

Our observable universe is estimated to have about, 1 billion trillion stars. The Milky Way galaxy has more than 100 billion starts its self. Our star is the host of earth. The only planet with intelligent, evolved life as far as we know officially.

Finding a planet with any type of lifeforms is like finding a needle in a haystack. The search for other earthlike planets is continuously ongoing. Scientists around the world use huge telescopes to look through the skies, decoding the stars themselves.

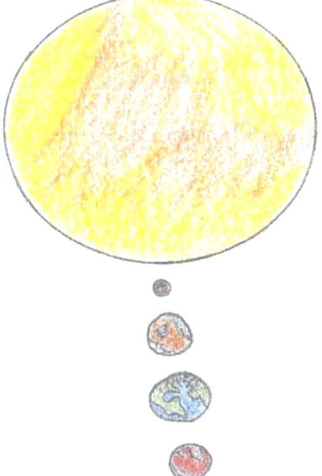

The earth is far enough from the sun to be habitable. Scientists are looking for other planets that share the same habitable zone. However, even if a planet is in a habitable zone it does not mean it sustains life, because there are many other factors involved to do so. Earth has all of these factors in check.

Some important factors of a life sustaining planet consist of: Distance from its sun, temperature, if liquid water is present, and atmospheric conditions for a breathable biome. A life sustaining planet needs all of these at the very least and they all need to be near perfect.

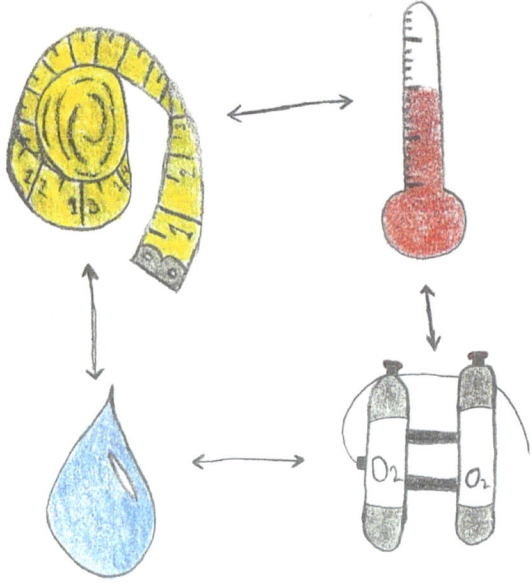

For a planet to sustain life the conditions need to be just right. A planet too close to its sun could take on too much radiation from possible solar flares. If it's too far away it could be too cold. Atmospheric and weather conditions need to be sustainable for any kind of organisms to breath and be alive.

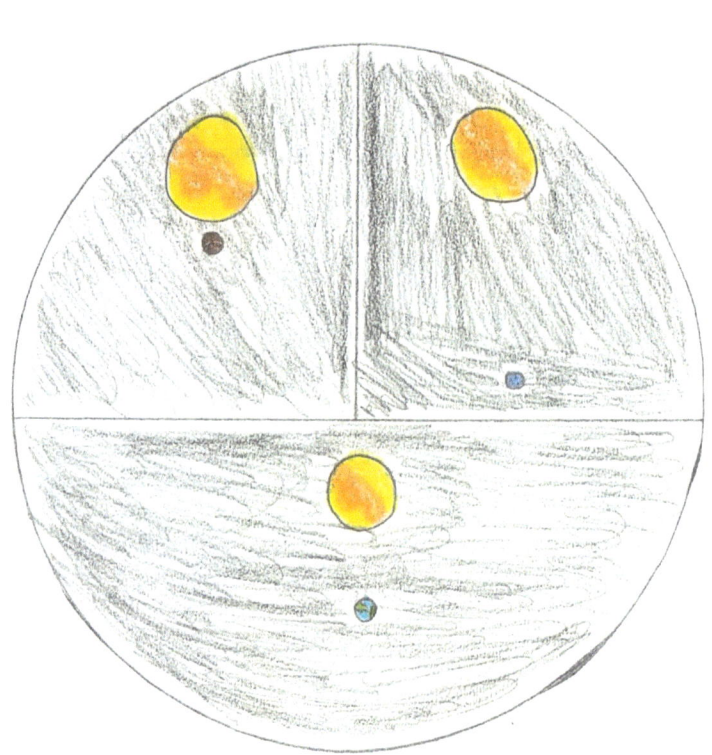

These planets are very hard to find, and questions still stand. Are they also sustainable for human life? If there are other lifeforms in the universe will we ever know of their presence officially? Will we find them or have they found us? Is all the speculation real or is it really science fiction? What kind of lifeforms are they? Etc.

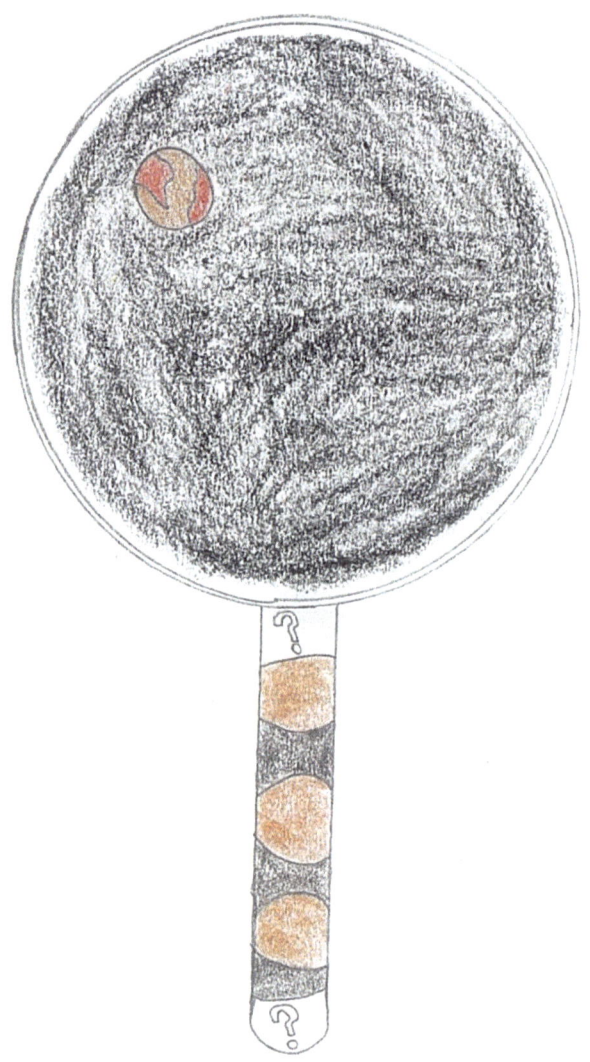

If there are other lifeforms in the universe, what would they be like? Would they be intelligent and evolved to their planet like us, or just microbes on a cellular level? Then of course there's the speculation of super intelligent lifeforms traveling through space with technology far ahead of ours.

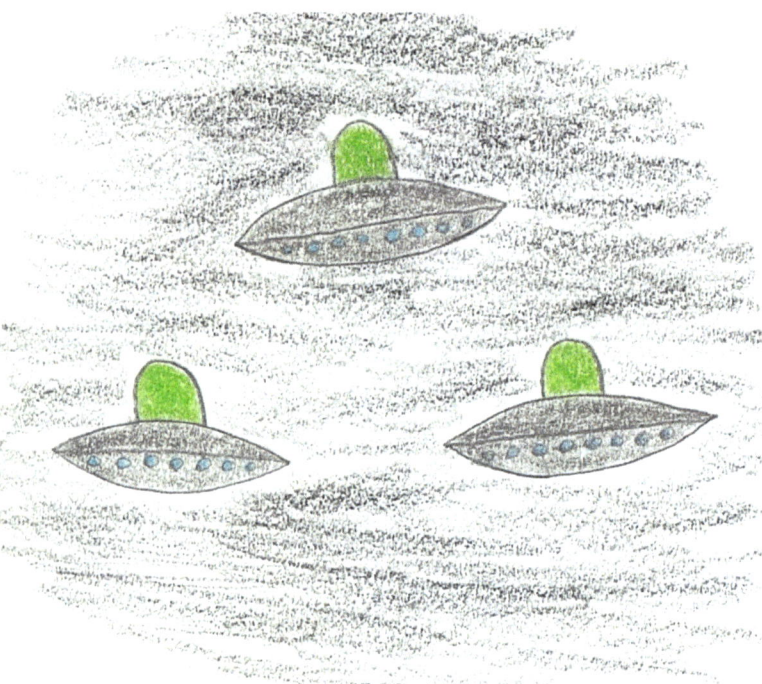

Curiosity about space and stars has always been a part of the human perspective. Maybe one day humanity will span across the stars and other lifeforms will truly be discovered. We could learn one day that our solar system is just one of many like it, and that there are many other stars of life.

Questioning the Star of Life is a booklet about our solar system and the universe its self. Challenging ideas of life on other planets and other worldly lifeforms in general.